Integrating Instruments of Power and Influence in National Security

Starting the Dialogue

Robert E. Hunter, Khalid Nadiri

Prepared for Carnegie Corporation of New York

RAND NATIONAL SECURITY RESEARCH DIVISION

The research described in this report was sponsored primarily by a grant from Carnegie Corporation of New York and was conducted under the auspices of International Programs within the RAND National Security Research Division (NSRD). NSRD conducts research and analysis for the Office of the Secretary of Defense, the Joint Staff, the Unified Commands, the defense agencies, the Department of the Navy, the Marine Corps, the U.S. Coast Guard, the U.S. Intelligence Community, allied foreign governments, and foundations.

ISBN 978-0-8330-4024-4

The RAND Corporation is a nonprofit research organization providing objective analysis and effective solutions that address the challenges facing the public and private sectors around the world. RAND's publications do not necessarily reflect the opinions of its research clients and sponsors.

RAND® is a registered trademark.

This publication was made possible by a grant from Carnegie Corporation of New York. The statements made and views expressed are solely the responsibility of the authors.

Published 2006 by the RAND Corporation
1776 Main Street, P.O. Box 2138, Santa Monica, CA 90407-2138
1200 South Hayes Street, Arlington, VA 22202-5050
4570 Fifth Avenue, Suite 600, Pittsburgh, PA 15213-2665
RAND URL: http://www.rand.org/
To order RAND documents or to obtain additional information, contact
Distribution Services: Telephone: (310) 451-7002;
Fax: (310) 451-6915; Email: order@rand.org

PREFACE

For at least a decade, it has become increasingly obvious that the deployment of U.S. military forces abroad requires much more than *military* activity. In many situations, the use—or threat of use—of military power must take place in close coordination and cooperation with non-military instruments of power and influence. This was demonstrated conclusively in both Bosnia and Kosovo, where NATO-led forces helped to keep the peace, but much of the nation-building work was done by non-military agencies and organizations. Some of the humanitarian relief and reconstruction work has been done by Civil-Military Cooperation units, but much was done by the U.S. State Department, the U.S. Agency for International Development (USAID)—and, indeed, by organizations like the United Nations, the Organization for Security and Co-operation in Europe (OSCE), the European Union, and a host of strictly non-governmental organizations (NGOs), many with purposes and mandates remote from the use of military force. With Afghanistan, Iraq, Darfur, earthquake relief in Pakistan, and tsunami relief in South and Southeast Asia, the need for military and non-military instruments of U.S. power and influence to work together in a systematic way—and to work with others, both non-American and non-governmental—has become the rule rather than the exception.

As a result of these developments and new requirements, the RAND Corporation, in cooperation with the American Academy of Diplomacy (AAD), is conducting a 15-month project on Integrating Instruments of Power and Influence in National Security, funded in part through a grant from the Rockefeller Brothers Fund. This project responds to a clear need for new approaches, on the part not just of the U.S. government—civilian and military—but also the private sector and NGOs. The inspiration for the project derives in part from a recognition that U.S. personnel "on the ground" have learned the lessons of working effectively together (among representatives of different U.S. agencies) and also with non-U.S. entities and even non-official entities. Every U.S. combatant commander who has deployed forces—or who has plans for deploying forces—has already learned these lessons, and today, from the Balkans to Southwest Asia, "integrating instruments of power and influence" has become daily business.

The RAND-AAD project is bringing together about 20 former U.S. officials with long and deep experience in these matters, to cull "best practices" from what Americans and others are doing abroad. Their goal is to see what common lessons have been learned and then to present findings and recommendations to help shape the way the U.S. government does business at the Washington end. This is thus a "bottom up" rather than "top down" perspective, one that will be critical in enabling the United States to be effective abroad in the years to come. At the same time, this project panel will include people from outside the United States—Canada and Europe, including NATO and the European Union—who have similar experience, plus a few individuals from the private sector, NGOs operating in the field, and some media and political observers. The project panel will meet several times in the next year, leading to publication and presentations in late 2007.

The project will be co-chaired by Ambassador Robert E. Hunter, RAND Senior Advisor and former U.S. ambassador to NATO; Ambassador Edward Gnehm, Jr., former U.S. ambassador to Jordan and Assistant Secretary of Defense; and a senior retired U.S. military officer.

The project was launched informally when, on March 21, 2006, the American Academy for Diplomacy and the American University School of International Service, in partnership with the RAND Corporation, convened a one-day conference devoted to a dialogue on "Integrating Instruments of Power and Influence in National Security." Two panels of current and former senior military officials and diplomats offered distinct perspectives, followed by a keynote speech by Ambassador Hunter. The program is attached. This report summarizes the main results of this conference.

This research was conducted within the International Security and Defense Policy Center of the RAND National Security Research Division (NSRD). NSRD conducts research and analysis for the Office of the Secretary of Defense, the Joint Staff, the Unified Combatant Commands, the defense agencies, the Department of the Navy, the Marine Corps, the U.S. Coast Guard, the U.S. Intelligence Community, allied foreign governments, and foundations. Publication of the report was made possible by a grant from Carnegie Corporation of New York.

For more information on RAND's International Security and Defense Policy Center, contact the director, James Dobbins. He can be reached by email at james_dobbins@rand.org; by phone at 703-413-1100, extension 5134; or by mail at the RAND Corporation, 1200 S. Hayes Street, Arlington, Virginia 20016. More information about RAND is available at www.rand.org.

CONTENTS

ACKNOWLEDGMENTS

We would like to express our deep gratitude to the American Academy of Diplomacy and the American University School of International Service for sponsoring this conference, particularly Dr. Louis Goodman, Ambassador L. Bruce Laingen, Professor Robert Pastor, the Honorable William Harrop, and Dr. Hoang Anh-Lam.

We would also like to thank the distinguished conference participants for their valuable contributions: General Joseph Hoar, USMC (ret.); Ambassador Anthony Quainton; Admiral Harold Gehman, USN (ret.); Lieutenant General Ray Odierno, USA; Ambassador Edward Gnehm; Ambassador Charles Freeman; Ambassador James Dobbins; and Ambassador Thomas E. McNamara. Finally, at RAND, we are grateful to Seth Jones and Nathan Chandler for their roles in making this report possible.

INTRODUCTION

Since the end of the Cold War, and particularly after September 11, 2001, the United States has been faced with an increasing number of complex challenges in which the effective means for securing U.S. interests is requiring new forms of interaction among a variety of instruments for projecting power and promoting influence. These interactions among U.S. military and civilian agencies, foreign allied governments, non-governmental organizations, the private sector, and multinational bodies have been most extensive at the field level, but have largely occurred without clear, precise, and comprehensive direction from the senior leadership in Washington. How this field-level cooperation occurs is valuable and informative in evaluating obstacles to, and making recommendations for, integration of U.S. instruments of power at the senior level. This report highlights those obstacles, evaluates possibilities, surveys "best practices," and makes recommendations, as discussed by the presenters and participants at the March 21, 2006, conference.

THE EVOLVING NATURE OF CONFLICTS

With the end of the Cold War came the fragmentation of particular ideologies, the acceleration of others, and an abrupt stop to political and economic flows that, in many parts of the world, had kept deep-seated sources of conflict from progressing. Political Islam, ethnic intolerance, and poverty became more visible as sources of conflict, particularly for many of the conflicts in which the United States has been engaged. When violent, these conflicts have included elements of insurgency and guerrilla warfare and have emerged in both urban and rural environments. Furthermore, in many cases there is no longer a bright line separating war from peace. Increasingly, it is essential to try shaping an environment so that conflict does not emerge (or, for example, so that the root causes of terrorism can be attacked), or to stabilize a country or region following the formal cessation of hostilities—what is sometimes called "nation building." In effect, a more or less seamless web of relationships and activities is developing that extends from the beginning of a crisis to its final resolution, a process that may take years or even decades. And the instruments involved, even in conflict situations, may not be primarily military. At the least, they may need to be coordinated with military activities of one form or another.

All the conference participants recognized that countering newly emergent threats before, during, and after the initiation of violence (where violence cannot be avoided) requires coordinated political, economic, military, human rights, and cultural activities, among others. But they noted that

_____

Rapporteur for the conference and principal drafter of this report was Khalid Nadiri of the RAND Corporation.

some key lessons of so-called nation building learned from U.S. foreign engagements at the field level have been lost on senior policymakers in Washington, precluding the planned cooperation so frequently improvised on the ground. Some participants attributed this widespread phenomenon to a certain "calculated ignorance," to the tendency for successive U.S. administrations to want to differentiate major aspects of foreign policy from that of prior administrations. In doing so, each successive administration ignores the substantial stock of experience and knowledge the United States has accumulated at the field and practical level, only to "relearn" after mistakes have been made. Many of the participants strongly asserted that it is critical that senior officials better understand the endurance of the new types of conflict the United States and other states are now facing, and work toward preserving lessons learned from them for the benefit of successive administrations.

SKILL ALLOCATION AND PERSONNEL COOPERATION

The Goldwater-Nichols Department of Defense Reorganization Act of 1986 has been highly successful in reorganizing the U.S. military into a more streamlined chain of command running from the President to the Secretary of Defense to Unified Combatant Commanders assigned to a discrete function of military operations (e.g., special operations, transportation, etc.) or a part of the world (e.g., Central Command, Southern Command). By restructuring military operations under a single sequence of authority, and by also requiring officers to work in interservice billets in order to advance to senior rank and assignments, the Goldwater-Nichols Act increased interservice cooperation and channeled competition into more productive results than in the past.

Poor cooperation still afflicts the military-civilian relationship today. Several conference participants stated that many personnel in U.S. military and civilian agencies are poorly qualified for their positions, particularly for senior or specialized posts. Further, they described a military-civilian relationship that is irregular and improvised because of dissonant bureaucratic cultures, perspectives of foreign engagement, and allocation of resources that does not reflect overall needs for getting the job done. Legislation similar to the Goldwater-Nichols Act applied to the various U.S. military and civilian agencies would significantly improve cooperation between civilians and the military and would encourage civilians to engage in activities that are essential but that, under current career patterns, may be unattractive.

Many participants argued that, by establishing measurable eligibility requirements, such as education and years of experience, this civilian Goldwater-Nichols-type legislation would instill rigor and reliability in the hiring and advancement process of people for senior or specialized positions.

For example, personnel with significant responsibilities in some civilian agencies are frequently underqualified for positions that demand a great deal of expertise, prior experience, and institutional contacts, and there is generally an increased need for professionalism in the Department of State and other civilian agencies with respect to tasks that entail close cooperation with the U.S. military. To achieve this professionalism, legislation would be needed so that junior officials could be groomed for leadership positions in interagency activity and experts could be insulated from reshuffling during political transitions and the resulting calculated ignorance described above. Further, foreign area specialists, who are routinely passed over in favor of nonspecialists for senior posts specific to a particular region, would be afforded extra consideration on the basis of their skill set or, alternatively, provided a separate structure in which to rise. Other notable recommendations included the creation of military counselors to advise the regional Assistant Secretaries of State and the increased empowerment of the Department of State's Office of Reconstruction and Stabilization (ORS). The ORS, which plans and executes civilian reconstruction and stabilization efforts in post-conflict states, has inadequate resources and too few qualified personnel, in contrast to its Canadian counterpart, CANADEM.

Additionally, legislation suggested by participants would establish collaborative and coordinated protocols, decisionmaking, and execution of operations among a broad group of military and civilian agencies. As suggested by the conference participants, this might include coordinated travel arrangements of senior officials in the military and civilian agencies, regular meetings of senior-level policymakers from multiple agencies in Washington, and frequent consultation and coordination among field personnel belonging to military and civilian agencies. For example, it was suggested that Assistant Secretaries of State for a particular region should meet regularly and officially with combatant commanders assigned to that same region, along with traveling on the same airplanes. It was also noted, however, that current public laws intended to prevent an overlap of authority by oversight committees obstruct this type of cooperation. These would need to be amended in some form.

But some obstacles to personnel cooperation cannot be entirely overcome by legislation. Participants from both the military and diplomatic panels described a dissonance in values, perspectives, and modes of deliberation between the U.S. military and civilian agencies, particularly with respect to the execution of policy. This, they argued, originates in Washington and resonates throughout the field, hindering both cooperation between policymakers at home and interagency coordination in foreign theaters. Strong leadership, they noted, would be required to push for cooperation between culturally distinct bureaucracies disinclined to cooperate.

FOREIGN CULTURES AND HISTORIES

A thorough understanding of foreign cultures and histories, particularly those of the countries in the Middle East and North Africa, is critical to appreciating the blended, multiform approach required to meet many of the challenges facing the United States today. Many conference participants remarked on the profound lack of understanding on the part of senior officials regarding the respective cultures and history of countries where the United States has been recently engaged, notably Afghanistan, Haiti, Iraq, and Colombia, thus detracting from their ability to form strategies consisting of the appropriate combination of political, economic, and military activities, among others. In Colombia, for example, senior officials in Washington pushed for the eradication of the coca crop but undervalued other difficulties in solving the country's drug problem, including the inhibitions on strengthening the rule of law and a lack of alternative livelihoods. Most recently, high-ranking U.S. government officials failed to recognize the ease with which ethnic and religious differences in Iraq could be exploited by insurgents. Most, if not all, field-level personnel observed and understood these distinctions through their daily activities, but policy prescriptions arriving from Washington often failed to take account of these realities. Better-informed leadership could help mitigate these problems.

RE-BALANCING

Both the military and diplomatic panels at the conference concluded that the civilian agencies are best equipped to undertake coordinated international non-military activities but are consistently denied the funding and authorization to do so. In contrast, the U.S. military receives much more money than the civilian agencies and is often charged with executing the type of activities for which USAID, the Department of State, or the Department of Justice is best prepared. A telling example was the 1994 deployment of the U.S. military in Haiti. Because of the relative lack of civilian agency personnel and the serious political, economic, and humanitarian deficiencies in Haiti, U.S. military personnel were required to undertake tasks with which they had scant familiarity, including reconstruction, public health, and rule of law projects. Participants recognized that even more familiar functions, such as the training of police, are best handled by the Department of State. These civilian agencies, the participants acknowledged, deserve more money and greater authority in executing non-military activities in cooperation with the established functions of the U.S. military. To accomplish this, Congress needs to be encouraged to move resources between the military and the civilian agencies. It lacks the full appreciation of the relative competencies of the civilian agencies necessary to empower them.

MULTILATERALS, NON-GOVERNMENTAL ORGANIZATIONS, AND COALITIONS

Conference participants acknowledged the usefulness of foreign governments and militaries, multilateral and multinational organizations, and NGOs at the field level. They noted that, despite significant differences in the U.S. approach to foreign policy among different agencies at the senior level, there is a great deal of agreement on short- and intermediate-term objectives at the field level—notably in the areas of security, reconstruction, justice, human rights, and democratization. But in return for the valuable counsel, expertise, legitimacy, or material support that can be gained from these external entities, the United States must be willing to share decisionmaking and influence with them, particularly with allied foreign governments and international agencies, such as the United Nations, NATO allies, and the European Union.

CONCLUSION

Out of this discussion of coordination and integration of instruments of power and influence emerged two central themes: utility and balance. Many of the obstacles to integration are the product of either deficiencies in knowledge, outmoded practices, or bureaucratic parochialism. These obstacles and future recommendations related to them need to be evaluated on the basis of utility, of the overall contribution to integration of U.S. national security. Further, a balance of human, financial, and jurisdictional resources needs to underlie any evaluation of integration. Effective cooperation between U.S. instruments of power and influence, as well as cooperation with non-U.S. or non-official actors, can only occur between agencies and other entities with comparable bargaining power.

APPENDIX: CONFERENCE PROGRAM

**The American Academy of Diplomacy &
The American University School of International Service**

**INTEGRATING INSTRUMENTS OF
POWER AND INFLUENCE IN NATIONAL SECURITY**

March 21, 2006
American University, Mary Graydon Center
Butler Board Room

Welcoming Remarks

Ambassador L. Bruce Laingen, President, American Academy of Diplomacy
Dr. Louis Goodman, Dean, School of International Service

Panel 1: The Military Perspective

Moderator: Anthony Quainton, Distinguished Diplomat in Residence, American University

Panelists: General Joseph Hoar, USMC (ret.), former Commander in Chief, U.S. Central Command

Admiral Harold Gehman, USN (ret.), former Supreme Allied Commander, Atlantic Command

Lieutenant General Ray Odierno, USA, Office of the Chairman, Joint Chiefs of Staff

Panel 2: The Diplomatic Perspective

Moderator: Ambassador Edward Gnehm, former Ambassador to Jordan, Kuwait, and Australia; Deputy Permanent Representative to the UN

Panelists: Ambassador James Dobbins, former Assistant Secretary of State for European Affairs and Special Envoy for Afghanistan

Ambassador Charles Freeman, former Ambassador to China and Saudi Arabia

Ambassador Thomas E. McNamara, former Ambassador to Colombia and former Assistant Secretary of State for Political/Military Affairs

Luncheon: Dr. Robert Hunter, former Ambassador to NATO